George Catlin

The Breath of Life

Mal-respiration, and its effects upon the enjoyments & life of man

George Catlin

The Breath of Life
Mal-respiration, and its effects upon the enjoyments & life of man

ISBN/EAN: 9783337182038

Printed in Europe, USA, Canada, Australia, Japan

Cover: Foto ©berggeist007 / pixelio.de

More available books at **www.hansebooks.com**

BREATH *of* LIFE

or

mal-Respiration.

and its

effects upon the enjoyments & life of man.

By

Geo Catlin

Author of " Notes of Travels amongst the North Amn. Indians."
&c., &c., &c.

LONDON·

TRUBNER & CO., 60 PATERNOSTER ROW.

1864.

Preface.

No perſon on Earth who reads this little work will condemn it: it is only a queſtion how many millions may look through it and benefit themſelves by adopting its precepts.

THE AUTHOR.

Entered according to Act of Congress, in the year 1861, by
JOHN WILEY,
In the Clerk ; Office of the District Court of the United States for the Southern District of New York.

R. CRAIGHEAD,
Printer, Stereotyper, and Electrotyper,
Carton Building,
81. 83, and 85 Centre Street.

BREATH OF LIFE.

THIS communication being made in the confident belief that very many of its Readers may draw from it hints of the higheſt importance to the enjoyment and prolongation of their lives, requires no other apology for its appearance, nor detention of the Reader from the information which it is deſigned to convey.

With the reading portion of the world it is generally known that I have devoted the greater part of my life in viſiting, and recording the looks of, the various native Races of North and South America; and during thoſe reſearches, obſerving the healthy condition and phyſical perfection of thoſe people, in their primitive ſtate, as contraſted with the deplorable mortality, the numerous diſeaſes and deformities, in civilized communities, I have been led to ſearch for, and able, I believe, to diſcover, the main cauſes leading to ſuch different reſults.

During my Ethnographic labours amongſt thoſe wild people I have viſited 150 Tribes, containing more than two millions of ſouls; and therefore have had, in all probability, more extenſive opportunities than any other man living, of examining their ſanitary ſyſtem; and if from thoſe examinations I have arrived at reſults of importance to the health and exiſtence of mankind, I fhall have achieved a double objeƈ in a devoted and toilfome life, and fhall enjoy a twofold fatisfaƈion in making them known to the world; and particularly to the Medical Faculty, who may perhaps turn them to good account.*

Man is known to be the moft perfeƈly conftructed of all the animals, and confequently he can endure more: he can out-travel the Horfe, the Dog, the Ox, or any other animal; he can faft longer; his natural life is faid to be "three fcore and ten years," while its *real, average length*, in civilized communities, is but half equal to that óf the brutes whofe natural term is not one third as long!

This enormous difproportion might be attributed to fome natural phyfical deficiency in the conftruƈion of Man, were it not that we find him in fome phafes of Savage life, enjoying almoft equal exemption. from difeafe and premature death, as the Brute creations; leading us to the irrefiftible conclufion that there is fome lamentable fault yet overlooked in the fanitary economy of civilized life.

The human Race and the various brute-fpecies have alike been created for certain refpeƈive terms of exiftence, and wifely fupplied with the phyfical means of fupporting that exiftence to its intended and natural end; and with the two creations, thefe powers would alike anfwer, as intended, for the whole term

---

* As the information contained in this little work is believed to be of equal importance to all claffes of fociety, and of all Nations, the Author has endcavoured to render it in the fimpleft poffible form, free from ambiguity of expreffion and profeffional technicality of language, that all may be able alike to appreciate it; and if the work contains feveral brief repetitions, they are only thofe which were *intended*, and fuch as always allowed, and even difficult to be avoided, in convey-ing important advice.

of natural life, except from some hereditary deficiency, or some practised abuse.

The horse, the dog, the ox, and others of the brute creations, we are assured by the breeders of those animals, are but little subject to the fatal diseases of the lungs and others of the respiratory or digestive organs; nor to diseases of the spine, to Idiocy or Deafness; and their teeth continuing to perform their intended functions to the close of natural life, not one in a hundred of these animals, with proper care and a sufficiency of food, would fail to reach that period, unless destroyed by intention or accident.

Mankind are everywhere a departure from this sanitary condition, though the Native Races oftentimes present a near approach to it, as I have witnessed amongst the Tribes of North and South America, amongst whom, in their *primitive condition*, the above-mentioned diseases are seldom heard of; and the almost unexceptional regularity, beauty, and soundness of their teeth last them to advanced life and old age.

In civilized communities, better sheltered, less exposed, and with the aid of the ablest professional skill, the sanitary condition of mankind, with its variety, its complication and fatality of diseases—its aches and pains, and mental and physical deformities, presents a more lamentable and mournful list, which plainly indicates the existence of some extraordinary, latent cause, not as yet sufficiently appreciated, and which it is the sole object of this little work to expose.

From the Bills of Mortality which are annually produced in the civilized world, we learn that in London and other large towns in England, and cities of the Continent, *on an average*, one half of the human Race die before they reach the age of five

years, and one half of the remainder die before they reach the
age of twenty-five, thus leaving but one in four to fhare the
chances of lafting from the age of twenty-five to old age.

Statiftical accounts fhowed not many years paft, that in
London, one half of the children died under *three* years, in
Stockholm, one half died under *two* years, and in Manchefter,
one half died under *five* years; but owing to recent improved
fanitary regulations, the numbers of premature deaths in thofe
cities are much diminifhed, leaving the average proportions as
firft given, no doubt, very near the truth, at the prefent time;
and ftill a lamentable ftatement for the contemplation of the
world, by which is feen the frightful gauntlet that civilized man
runs in his paffage through life.

The fanitary condition of the Savage Races of North and
South America, a few inftances of which I fhall give, not by
quoting a variety of authors, but from eftimates carefully made
by myfelf, whilft travelling among thofe people, will be found
to prefent a ftriking contraft to thofe juft mentioned, and fo
widely different as naturally, and very juftly, to raife the inquiry
into the caufes leading to fuch diffimilar refults.

Several very refpeftable and credible modern writers have
undertaken to fhow, from a hoft of authors, that premature
mortality is greater amongft the Savage, than amongft the Civilized Races; which is by no means true, excepting amongft
thofe communities of favages who have been corrupted, and
their fimple and temperate modes of life changed by the diffipations and vices introduced among them by civilized people.

In order to draw a fair contraft between the refults of habits
amongft the two races, it is neceffary to contemplate the two
people living in the uninvaded habits peculiar to each; and it

would be well alfo, for the writer who draws thofe contrafts, to fee with his own eyes, the cuftoms of the native Races, and obtain his information from the lips of the people themfelves, inftead of trufting to a long fucceffion of authorities, each of which has quoted from his predeceffor, when the original one has been unworthy of credit or has gained his information from unreliable, or ignorant, or malicious fources.

There is, perhaps, no other fubject upon which hiftorians and other writers are more liable to lead the world into erroneous conclufions than that of the true native cuftoms and charafter of Aboriginal Races; and that from the univerfal dread and fear which have generally deterred hiftorians and other men of Science from penetrating the folitudes inhabited by thefe people, in the practice of their primitive modes.

There always exifts a broad and moving barrier between favage and civilized communities, where the firft fhaking of hands and acquaintance take place, and over which the demoralizing and deadly effects of diffipation are taught and practifed; and from which, unfortunately, both for the charafter of the barbarous races and the benefit of fcience, the cuftoms and the perfonal appearance of the favage are gathered and portrayed to the world.

It has been too much upon this field that hiftorians and other writers have drawn for the exaggerated accounts which have been publifhed, of the exceffive mortality amongft the favage Races of America, leading the world to believe that the actual premature wafte of life caufed by the diffipations and vices introduced, with the accompanying changes in the modes of living in fuch diftricts, were the proper ftatiftics of thofe people.

I have vifited thefe femi-civilized degradations of favage
life in every degree of latitude in North America, and to a
great extent alfo in Central and South America, and as far as
this fyftem extends, I agree with thofe writers who have con-
tended in general terms, that premature mortality is proportion-
ally greater amongft the Native Races than in Civilized communities; but as I have alfo extended my vifits and my inquiries into the tribes in the fame latitudes, living in their primitive
ftate, and practifing their native modes, I offer myfelf as a
living witnefs, that whilft in that condition, the Native Races
in North and South America are a healthier people, and lefs
fubject to premature mortality (fave from the accidents of War
and the Chafe, and alfo from Small-pox and other peftilential
difeafes introduced amongft them), than any civilized Race in
exiftence.

Amongft a people who preferve no Records and gather no
Statiftics, it has been impoffible to obtain *exact* accounts of their
annual deaths, or ftrict proportionate eftimates of deaths before
and between certain ages ; but from verbal eftimates given me
by the Chiefs and Medical men in the various tribes, and whofe
ftatements may in general be relied on as very near the truth,
there is no doubt but I have been able to obtain information on
thefe points which may fafely be relied on as a juft average of
the premature mortality in many of thofe Tribes, and which
we have a right to believe would be found to be much the
fame in moft of the others.

As to the melancholy proportions of deaths of children in
civilized communities already given, there is certainly no parallel
to it to be found amongft the North or South American Tribes,
where they are living according to their primitive modes ; nor

do I believe that a fimilar mortality can be found amongft the children of any aboriginal race on any part of the globe.

Amongft the North American Indians, at all events, where two or three children are generally the utmoft refults of a marriage, fuch a rate of mortality could not exift without foon depopulating the country; and as a juftification of the general remark I have made, the few following inftances of the numerous eftimates which I received and recorded amongft the various tribes, I offer in the belief that they will be received as matters for aftonifhment, calling for fome explanation of the caufes of fo vide a contraft between the Bills of Mortality in the two Races.

Whilft refiding in a fmall village of Guarani of 250 perfons, on the banks of the Rio Trombutas, in Brazil; amongft the queftons which I put to the Chief, I defired to know as near as poffible, the number of children under 10 years of age, which his village had loft within the laft 10 years, a fpace of time over which his recollection could reach with tolerable accuracy. After he and his wife had talked the thing over for fome time, they together made the following reply, viz.—that "they could recolle& but three deaths of children within that fpace of time : one of thefe was drowned, a fecond one was killed by the kick of a hofe, and the third one was bitten by a Rattle Snake."

Thi fmall tribe, or Band, living near the bafe of the Acarai Mountains, refembled very much in their perfonal appearance and modes of life, the numerous bands around them; all mounted on good horfes; living in a country of great profufion, both of animal and vegetable food.

The '*Sleepy Eyes*," a celebrated chief of a Band of Sioux,

power, and a most tyrannical one, indeed, and though he helped to foster the ballets which won the chief delight of the grand monarch and his court, he composed twenty operas, some of the airs of which may still be studied with profit and heard with pleasure, and fixed the form of the French grand opera, which recognized then and still recognizes the keen instincts of the French people for the drama. Italian influences did not lose their hold in Paris, however, and when Gluck came, in the Eighteenth Century, to write in the manner that might have been expected to make an irresistible appeal to the French people, he had to fight his bitter battle with Piccini. In England, the principles represented by the Florentines found expression in a setting of a masque from Ben Jonson in 1617, by Nicolo Laniere, an Italian born in London; but the fashion of setting an entire stage play to music was not established by Laniere's experiment. Even when England's most powerful and original genius, Henry Purcell (1658-1695) came, the operatic form still lagged. Purcell was a pupil of Pelham Humphries, a pupil of Lully; yet Purcell, with unmistakable dramatic instincts, wrote no complete opera, but only incidental dramatic music for masques and plays, though some of these compositions have the form, dimensions and significance of operatic scenes. Italian opera of the accepted Italian type came into dominant vogue with Handel in 1711.

What was opera like at the close of the period which has now been outlined? I can only give a few significant hints and leave the filling out to the imagination of the reader, or the completion of his knowledge by further study. In Germany and England, we are confronted for a time with an anomaly of language. The purveyors felt that the people ought to understand the words of the play, but they were dependent on foreign singers and foreign composers to a great extent, and they knew that their own languages were not as well adapted to Italian music as the Italian. So, for a space, they made use of two languages, Italian and the vernacular. Handel's "Almira," written for Hamburg, has

German recitatives for the dialogue, and Italian arias. For three years in London, Italian and English were mixed in the manner amusingly described by Addison:

" The King or hero of the play generally spoke in Italian, and his slaves answered him in English; the lover frequently made his court and gained the heart of his princess in a language which she did not understand. At length the audience got tired of understanding half the opera and to ease themselves entirely of the fatigue of thinking, so ordered it that the whole opera was performed in an unknown tongue." Addison thought that the grandchildren of his generation would wonder at the conduct on the part of their forefathers, in listening to plays which they did not understand; but the English and American people do the same thing today.

But in Italy itself, where the language was understood, the opera was less artificial. At the outset the subjects had been classical; very naturally, indeed, the record starts with the story of Orpheus and Eurydice. Then they became antique—historical. But it made no difference whether the hero was a god, a demi-god, an ancient monarch, or a man of war. It was his business to run about the stage, generally in disguise, and sing elaborate tunes in an unsexed voice. A hard and fast formula governed the construction of operas down almost to the Mozart period, the period from which present, popular and practical knowledge may be said to date. The plot had to be classical; there had to be six characters and six only (three women and three men); occasionally a woman might take a man's part, but many of the men sang with women's voices; there were three acts and in each of the three each character sang an air; there were five varieties of airs, but each kind had the da capo; that is, after it had been finished the singer returned to the beginning and sang the first part over again, this time with such embellishments as he or she could invent. The various kinds of arias were designed to display the capacity of the singers in the sustained style, their ability to sustain long notes, to declaim the words rapidly and expressively, to sing long flourishes

(or hunch-back), of *Deaf and Dumb*, or of other deformity of a difabling kind.

· The inftances which I have thus far ftated, as *rather extraordinary cafes* of the healthfulnefs of their children, in the above tribes, are neverthelefs, not far different from many others which I have recorded in the numerous tribes which I have vifited; and the *apparently* fingular exemption of the Mandans, which I have mentioned, from mental and phyfical deformities, is by no means peculiar to that tribe; but, almoft without exception, is applicable to all the tribes of the American Continent, where they are living in their primitive condition, and according to their original modes.

This Tribe subfifts chiefly on Buffalo meat, and maize or Indian corn, which they raifed to a confiderable extent.

Amongft two millions of thefe wild people whom I have vifited, I never faw, or heard of a *hunch-back* (crooked fpine) though my inquiries were made in every tribe; nor did I ever *fee* an *Idiot* or *Lunatic* amongft them, though I heard of fome three or four, during my travels, and perhaps of as many *Deaf and Dumb*.*

---

* Some writers upon whom the world have relied for a correct account of the cuftoms of the American Indians, have affigned as the caufe of the almoft entire abfence of mental and phyfical deformities amongft thefe people, that they are in the habit of putting to death all who are thus afflicted; but fuch is not only an unfounded and unjuft, but difgraceful affumption on the part of thofe by whom the opinions of the world have been led; for, on the contrary, in every one of the few very cafes of the kind, which I have met or could hear of, amongft two millions of thefe people, thefe unfortunate creatures were not only fupplied and protected with extraordinary care and fympathy, but were in all cafes guarded with a

Shar-re-tar-ruſhe, an aged and venerable chief of the Pawnee-Piɛts, a powerful tribe living on the head-waters of the Arkanſas River, at the baſe of the Rocky Mountains, told me in anſwer to queſtions, " we very ſeldom loſe a ſmall child—none of our women have ever died in childbirth—they have no medical attendance on theſe occaſions—we have no idiots or lunatics—nor any Deaf and Dumb, or Hunch-backs, and our children never die in teething."

This Tribe I found living entirely in their primitive ſtate; their food, Buffalo fleſh and Maize, or Indian corn.

Ski-ſe-ro-ka, chief of the Kiowas, a ſmall Tribe, on the head-waters of the Red River, in Weſtern Texas, replied to me, " my wife and I have loſt two of our ſmall children, and perhaps ten or twelve have died in the tribe in the laſt ten years—we have loſt none of our children by teething—we have no Idiots, no Deaf and Dumb, nor hunch-backs."

This Tribe I found living in their primitive condition, their food Buffalo fleſh and venison.

Cler-mont, chief of the Oſages, replied. to my queſtions, "before my people began to uſe '*fire-water*,' it was a very unuſual thing for any of our women to loſe their children; but I am ſorry to ſay that we loſe a great many of them now; we have no Fools (Idiots), no Deaf and Dumb, and no hunch-backs—our women never die in childbirth nor have dead children."

Naw-kaw, chief of the Winnebagoes, in Wiſconſin, the

ſuperſtitious care, as the probable receptacles of ſome important myſtery, deſigned by the Great Spirit, for the undoubted benefit of the families or Tribes to which they belonged.

remnant of a numerous and warlike tribe, now ſemi-civilized and reduced, " our children are not now near ſo healthy as they were when I was a young man; it was then a very rare thing for a woman to loſe her child; now it is a very difficult thing to raiſe them,"—to which his wife added—" ſince our huſbands have taken to drink ſo much whiſkey our babies are not ſo ſtrong, and the greater portion of them die; we cannot keep them alive."   The chief continued, " we have no Idiots, no Deaf and Dumb, and no hunch-backs; our women never die in childbirth, and they do not allow Doctors to attend them on ſuch occaſions."

Food of this Tribe, Fiſh, veniſon, and vegetables.

Kee-mon-ſaw, chief of the *Kaſkaſkias*, on the Miſſouri, once a powerful and warlike tribe, told me that he could recollect when the children of his tribe were very numerous and very healthy, and they had then no Idiots, no deaf and dumb, nor hunch-backs; but that the ſmall-pox and whiſkey had killed off the men and women, and the children died very faſt.   " My Mother," ſaid he, " who is very old, and my little ſon and myſelf, all of whom are now before you, are all that are left in my tribe, and I am the chief!"

The above, which are but a very few of the numerous eſtimates which I have gathered, when compared with the ſtatiſtics of premature deaths and mental and phyſical deformities in civilized communities, form a contraſt ſo ſtriking, between the ſanitary conditions of the two Races who are born the ſame, and whoſe terms of natural life are intended to be equal, as plainly to ſhow, that through the vale of their exiſtence, in civilized Races, there muſt be ſome hidden cauſe of diſeaſe not

yet fufficiently appreciated, and which the *Materia Medica* has
not effectually reached.

Under this conviction I have been ftimulated to fearch
amongft the Savage Races for the caufes of their exemption
from, and amongft the civilized communities for the caufes of
their fubjection to, fo great a calamity, and this I believe I have
difcovered, commencing in the cradle, and accompanying
civilized mankind through the painful gauntlet of life to the
grave; and in poffeffion of this information, when I look into
the habits of fuch communities, and fee the operations of this
caufe, and its lamentable effects, I am not in the leaft aftonifhed
at the frightful refults which the lifts of mortality fhow; but it
is matter of furprife to me that they are not even more lamentable, and that Nature can fuccefsfully battle fo long as fhe does,
againft the abufes with which fhe often has to contend.

This caufe I believe to be the fimple neglect to fecure the
vital and intended advantages to be derived from quiet and
natural fleep; the great phyfician and reftorer of mankind, both
Savage and Civil, as well as of the Brute creations.

Man's cares and fatigues of the day become a daily difeafe,
for which quiet fleep is the cure; and the All-wife Creator has
fo conftructed him that his breathing lungs fupport him through
that fleep, like a perfect machine, regulating the digeftion of the
ftomach and the circulation of the blood, and carrying repofe
and reft to the utmoft extremity of every limb; and for the
protection and healthy working of this machine through the
hours of repofe, He has formed him with noftrils intended for
meafuring and tempering the air that feeds this moving principle and fountain of life; and in proportion as the quieting
and reftoring influence of the lungs in natural repofe, is carried

to every limb and every organ, fo in *unnatural* and *abufed* repofe,
do they fend their complaints to the extremities of the fyftem,
in various difeafes; and under continued abufe, fall to pieces
themfelves, carrying inevitable deftruction of the fabric with
them in their decay.

The two great and primary phafes in life and mutually
dependant on each other, are *waking* and *fleeping ;* and the abufe
of either is fure to interfere with the other. For the firft of
thefe there needs a lifetime of teaching and practice ; but for
the enjoyment of the latter, man needs no teaching, provided
the regulations of the All-wise Maker and Teacher can have
their way, and are not contravened by pernicious habits or
erroneous teaching.

If man's unconfcious exiftence for nearly one-third of the
hours of his breathing life depends from one moment to another,
upon the air that paffes through his noftrils; and his repofe
during thofe hours, and his bodily health and enjoyment between
them, depend upon the foothed and tempered character of the
currents that are paffed through his nofe to his lungs, how
myfterioufly intricate in its conftruction and important in its
functions is that feature, and how difaftrous may be the omiffion
in education which fanctions a departure from the full and
natural ufe of this wife arrangement?

When I have feen a poor Indian woman in the wildernefs,
lowering her infant from the breaft, and preffing its lips together
as it falls afleep in its cradle in the open air, and afterwards
looked into the Indian multitude for the refults of fuch a practice, I have faid to myfelf, " glorious education! fuch a Mother
deferves to be the nurfe of Emperors." And when I have feen
the *careful, tender mothers* in civilized life, covering the faces of

their infants fleeping in overheated rooms, with their little mouths open and gafping for breath; and afterwards looked into the multitude, I have been ftruck with the evident evil and lafting refults of this incipient ftage of education; and have been more forcibly ftruck, and fhocked, when I have looked into the Bills of Mortality, which I believe to be fo frightfully fwelled by the refults of this habit, thus contracted, and practifed in contravention to Nature's defign.

There is no animal in nature excepting Man, that fleeps with the mouth open; and with mankind, I believe the habit, which is not natural, is generally confined to civilized communities, where he is nurtured and raifed amidft enervating luxuries and unnatural warmth, where the habit is eafily contracted, but carried and practifed with great danger to life in different latitudes and different climates; and, in fudden changes of temperature, even in his own houfe.

The phyfical conformation of man alone affords fufficient proof that this is a habit againft inftinct, and that he was made, like the other animals, to fleep with his mouth fhut—fupplying the lungs with vital air through the noftrils, the natural channels; and a ftrong corroboration of this fact is to be met with amongft the North American Indians, who ftrictly adhere to Nature's law in this refpect, and fhow the beneficial refults in their fine and manly forms, and exemption from mental and phyfical difeafes, as has been ftated.

The Savage infant, like the offspring of the brute, breathing the natural and wholefome air, generally from inftinct, clofes its mouth during its fleep; and in all cafes of exception the mother rigidly (and *cruelly*, if neceffary) enforces Nature's Law in the manner explained, until the habit is fixed for life, of the

2

importance of which she seems to be perfectly well aware. But when we turn to civilized life, with all its comforts, its luxuries, its science, and its Medical skill, our pity is enlisted for the tender germs of humanity, brought forth and caressed in smothered atmospheres which they can only breathe with their mouths wide open, and nurtured with too much thoughtlessness to prevent their contracting a habit which is to shorten their days with the croup in infancy, or to turn their brains to Idiocy or Lunacy, and their spines to curvatures—or in manhood, their sleep to fatigue and the nightmare, and their lungs and their lives to premature decay.*

If the habit of sleeping with the mouth open is so destructive to the human constitution, and is caused by sleeping in confined and overheated air, and this under the imprudent sanction of mothers, they become the primary causes of the misery of their own offspring; and to them, chiefly, the world must look for the correction of the error, and, consequently, the benefaction of mankind. They should first be made acquainted with the fact that their infants don't require heated air, and that

---

* The weekly Bills of Mortality in London show an amount of 10, 15, and sometimes 20 deaths of infants per week, from suffocation, in bed with their parents; and Mr. Wakley, in May, 1860, in an inquest on an infant, stated that " he had held inquests over more than 100 Infants which had died during the past winter, from the same cause, their parents covering them entirely over, compelling them to breathe their own breath."—*Times.*

The Registrar General shows an average of over 700,000 infants born in England per annum, and over 100,000, which die under one year of age—12,738 of these of Bronchitis, 3,660 from the pains of teething, and 19,000 of convulsions, and says, " suffocation in bed, by overlaying or shutting off the air from the child, is the most frequent cause of violent deaths of children in England."

they had better fleep with their heads out of the window than under their mother's arms—that middle-aged and old people require more warmth than children, and that to embrace their infants in their arms in their fleep during the night, is to fubject them to the heat of their own bodies; added to that of feather beds and overheated rooms, the relaxing effects of which have been mentioned, with their pitiable and fatal confequences.

There are many, of courfe, in all ranks and grades of fociety, who efcape from contracting this early and dangerous habit, and others who commence it in childhood, or in manhood, a very few of whom live and fuffer under it to old age, with conftitutions fufficiently ftrong to fupport Nature in her defperate and continuous ftruggle againft abufe.

When we obferve amongft very aged perfons that they almoft uniformly clofe the mouth firmly, we are regarding the refults of a long practifed and healthy habit, and the furviving few who have thereby efcaped the fatal confequences of the evil practice I am condemning.

Though the majority of civilized people are more or lefs addicted to the habit I am fpeaking of, comparatively few will admit that they are fubject to it. They go to fleep and awake, with their mouths fhut, not knowing that the infidious enemy, like the deadly Vampire that imperceptibly fucks the blood, gently fteals upon them in their fleep and does its work of death whilft they are unconfcious of the evil.

Few people can be convinced that they fnore in their fleep, for the fnoring is ftopped when they awake; and fo with breathing through the mouth, which is generally the caufe of fnoring— the moment that confcioufnefs arrives the mouth is clofed, and Nature refumes her ufual courfe.

In natural and refreshing sleep, man breathes but little air; his pulse is low; and in the most perfect state of repose he almost ceases to exist. This is neceffary, and most wisely ordered, that his lungs, as well as his limbs, may rest from the labour and excitements of the day.

· Too much sleep is often said to be destructive to health; but very few persons will sleep too much for their health, provided they sleep in the right way. Unnatural sleep, which is irritating to the lungs and the nervous syftem, fails to afford that rest which sleep was intended to give, and the longer one lies in it, the less will be the enjoyment and length of his life. Any one waking in the morning at his usual hour of. rising, and finding by the drynefs of his mouth, that he has been sleeping with the mouth open, feels fatigued, and a wish to go to sleep again; and, convinced that his rest has not been good, he is ready to admit the truth of the statement above made.

There is no perfect sleep for man or brute, with the mouth open; it is unnatural, and a strain upon the lungs which the expreffion of the countenance and the nervous excitement plainly show.

Lambs, which are nearly as tender as human infants, commence immediately after they are born, to breathe the chilling air of March and April, both night and day, afleep and awake, which they are able to do, becaufe they breathe it in the way that Nature defigned them to breathe. New-born infants in the Savage Tribes are expofed to nearly the fame neceffity, which they endure perfectly well, and there is no reafon why the oppofite extreme should be practifed in the civilized world, entailing fo much misfortune and mifery on mankind.

It is a pity that at the very *starting point* of life—Man should be ftarted wrong—that mothers fhould be under the erroneous belief that while their infants are awake they muft be watched; but afleep, they are "doing well enough."

Education is twofold, mental and phyfical; the latter of which alone, at this early ftage, can be commenced; and the mother fhould know that fleep, which is the great renovater and regulator of health, and in fact, the *food* of *life*, fhould be enjoyed in the manner which Nature has defigned; and therefore that her clofeft fcrutiny and watchfulnefs, like that of the poor Indian woman, fhould guard her infant in thofe important hours, when the fhooting germs of conftitution are ftarting, on which are to depend the happinefs or mifery of her offspring.

It requires no more than common fenfe to perceive that Mankind, like all the Brute creations, fhould clofe their mouths when they clofe their eyes in fleep, and breathe through their noftrils, which were evidently made for that purpofe, inftead of dropping the under jaw and drawing an over draught of cold air directly on the lungs, through the mouth; and that in the middle of the night, when the fires have gone down and the air is at its coldeft temperature—the fyftem at reft, and the lungs the leaft able to withftand the fhock.

For thofe who have fuffered with weaknefs of the lungs or other difeafes of the cheft, there needs no proof of this fact; and of thofe, if any, who are yet incredulous, it only requires that they fhould take a candle in their hand and look at their friends afleep and fnoring; or with the Nightmare (or without it), with their eyes fhut and their mouths wide open—the very pictures of diftrefs—of fuffering, of Idiocy, and Death; when Nature defigned that they fhould be fmiling in the foothing and invigorating forgetfulnefs of the fatigues and anxieties of the

day, which are diffolving into pleafurable and dreamy fhadows
of "realities gone by."

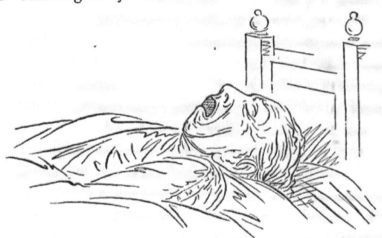

Who ever waked out of a fit of the Nightmare in the middle of the night with his mouth ftrained open and dried to a
hufk, not knowing when, or from where, the faliva was coming

to moiften it again, without being willing to admit the mifchief
that fuch a habit might be doing to the lungs, and confequently

to the ftomach, the brain, the nerves, and every other organ of the fyftem?

Who, like myfelf, has fuffered from boyhood to middle age, everything but death from this enervating and unnatural habit, and then, by a *determined* and *uncompromifing* effort, has thrown it off, and gained, as it were, a new leafe of life and the enjoyment of reft—which have lafted him to an advanced age through all expofures and privations, without admitting the mifchief of its confequences?

Nothing is more certain than that for the prefervation of human health and life, that moft myfterious and incomprehenfible, felf-acting principle of life which fupports us through the reftoring and unconfcious vale of fleep, fhould be protected and aided in every way which Nature has prepared for the purpofe, and not abufed and deranged by forcing the means of its fupport through a different channel.

We are told that "the breath of life was breathed into man's noftrils"—then why fhould he not *continue* to live by breathing it in the fame manner?*

---

* A recently invented aid for the lungs, which the ufual efforts for pecuniary refults, and the accuftomed and unfortunate rage for novelties have been pufhing into extenfive ufe, has been doing great mifchief in fociety during the laft few years; and by its injudicious ufe, I believe thoufands on thoufands have been hurried to the grave. I refer to the "Refpirators," fo extenfively in ufe, and as generally "in fafhion," amongft the Fair Sex. For perfons very weak in the lungs, and who have contracted the habit fo ftrong and fo long that they cannot breathe excepting through the open mouth, this appliance may be beneficial, in the open air; but thoufands of others, to be eccentric or fafhionable, place it over their mouths when they ftep into the ftreet; and to make any ufe of it, muft open their mouths and breathe through it, by which indifcretion they are thought-

The mouth of man, as well as that of the brutes, was made for the reception and maſtication of food for the ſtomach, and other purpoſes; but the noſtrils, with their delicate and fibrous linings for purifying and warming the air in its paſſage, have been myſteriouſly conſtructed, and deſigned to ſtand guard over the lungs—to meaſure the air and equalize its draughts, during the hours of repoſe.

The atmoſphere is nowhere pure enough for man's breathing until it has paſſed this myſterious refining proceſs; and therefore the imprudence and danger of admitting it in an unnatural way, in double quantities, upon the lungs, and charged with the ſurrounding epidemic or contagious infections of the moment.

The impurities of the air which are arreſted by the intricate organizations and mucus in the noſe are thrown out again from its interior barriers by the returning breath; and the tingling excitements of the few which paſs them, cauſe the muſcular involitions of ſneezing, by which they are violently and ſucceſs fully refiſted.

The air which enters the lungs is as different from that which enters the noſtrils as diſtilled water is different from the water in an ordinary ciſtern or a frog-pond. The arreſting and purifying proceſs of the noſe, upon the atmoſphere with its poiſonous ingredients, paſſing through it, though leſs perceptible, is not leſs diſtinct, nor leſs important than that of the mouth,

---

leſſly contracting the moſt dangerous habit which they can ſubject themſelves to, and oftentimes catching their death in a few days, or in a few hours; little aware that cloſed lips are the beſt protection againſt cold air, and their noſtrils the beſt and ſafeſt of all Reſpirators.

which stops cherry-stones and fish-bones from entering the stomach.

This intricate organization in the structure of man, unaccountable as it is, seems in a measure divested of mystery, when we find the same phenomena (and others perhaps even more surprising) in the physical conformation of the lower order of animals; and we are again more astonished when we see the mysterious sensitiveness of that organ instinctively and instantaneously separating the *gases*, as well as arresting and rejecting the *material* impurities of the atmosphere.

This unaccountable phenomenon is seen in many cases. We see the fish, surrounded with water, breathing the air upon which it exists. It is a known fact that man can inhale through his nose, for a certain time, *mephitic air*, in the bottom of a well, without harm; but if he opens his mouth to answer a question, or calls for help, in that position, his lungs are closed and he expires. Most animals are able to inhale the same for a considerable time without destruction of life, and, no doubt, solely from the fact that their respiration is through the nostrils, in which the poisonous effluvia are arrested.

There are many mineral and vegetable poisons also, which can be inhaled by the nose without harm, but if taken through the mouth destroy life. And so with poisonous reptiles, and poisonous animals. The man who kills the Rattlesnake, or the Copperhead, and stands alone over it, keeps his mouth shut, and receives no harm; but if he has companions with him, with whom he is conversing over the carcases of these reptiles, he inhales the poisonous effluvia through the mouth, and becomes deadly sick, and in some instances death ensues.

Infinitesimal insects also, not visible to the naked eye, are inhabiting every drop of water we drink and every breath of air we breathe; and minute particles of vegetable fubftances, as well as of poifonous minerals, and even glafs and filex, which float imperceptibly in the air, are difcovered coating the refpiratory organs of man; and the clafs of birds which catch their food in the air with open mouths as they fly, receive thefe things in quantities, even in the hollow of their bones, where they are carried and lodged by the currents of air, and detected by microfcopic inveftigation.

Againft the approach of thefe things to the lungs and to the eye, Nature has prepared the guard by the mucous and organic arrangements, calculated to arreft their progrefs. Were it not for the liquid in the eye, arrefting, neutralizing, and carrying out the particles of duft communicated through the atmofphere, Man would foon become blind; and but for the mucus in his noftrils, abforbing and carrying off the poifonous particles and effluvia for the protection of the lungs and the brain, mental derangement, confumption of the lungs, and death would enfue.

How eafy, and how reafonable, it is to fuppofe then, that the inhalation of fuch things to the lungs through the expanded mouth and throat may be a caufe of confumption and other fatal difeafes attaching to the refpiratory organs; and how fair a fuppofition alfo, that the deaths from the dreadful Epidemics, fuch as Cholera, Yellow Fever, and other peftilences, are caufed by the inhalation of animalcules in the infected diftricts; and that the victims to thofe difeafes are thofe portions of fociety who inhale the greateft quantities of thofe poifonous infects to the lungs and to the ftomach.

In man's waking hours, when his limbs, and mufcles, and

his mind, are all in action, there may be but little harm in inhaling through the mouth, if he be in a healthy atmofphere; and at moments of violent action and excitement, it may be neces-ary. But when he lies down at night to reft from the fatigues of the day, and yields his fyftem and all his energies to the repofe of fleep; and his volition and all his powers of refiftance are giving way to its quieting influence, if he gradually opens his mouth to its wideft ftrain, he lets the enemy in that chills his lungs—that racks his brain—that paralyfes his ftomach—that gives him the nightmare—brings him Imps and Fairies that dance before him during the night; and during the following day, headache—toothache—rheumatifm—dyfpepfia, and the gout.

That man knows not the pleafure of fleep; he rifes in the morning more fatigued than when he retired to reft—takes pills and remedies through the day, and renews his difeafe every night. A guilty confcience is even a better guarantee for peaceful reft than fuch a treatment of the lungs during the hours of fleep. Deftructive irritation of the nervous fyftem and inflammation of the lungs, with their confequences, are the immediate refults of this unnatural habit; and its continued and more remote effects, confumption of the lungs and death.

Befides this frequent and moft fatal of all difeafes, Bronchitis, Quinfey, Croup, Afthma, and other difeafes of the refpiratory organs, as well as Dyfpepfia, gout of the ftomach, Rickets, Diarrhœa, difeafes of the liver, the heart, the fpine, and the whole of the nervous fyftem, from the brain to the toes, may chiefly be attributed to this deadly and unnatural habit; and any Phyfician can eafily explain the manner in which thefe various parts of the fyftem are thus affected by the derangement of the natural functions of the machine that gives them life and motion.

All perfons going to fleep fhould think, not of their bufinefs, not of their riches or poverty, their pains or their pleafures, but, of what are of infinitely greater importance to them, their lungs; their beft friends, that have kept them alive through the day, and from whofe quiet and peaceful repofe they are to look for happinefs and ftrength during the toils of the following day. They fhould firft recollect that their natural food is frefh air; and next, that the channels prepared for the fupply of that food are the noftrils, which are fupplied with the means of purifying the food for the lungs, as the mouth is conftructed to felect and mafticate the food for the ftomach. The lungs fhould be put to reft as a fond Mother lulls her infant to fleep; they fhould be fupplied with vital air, and protected in the natural ufe of it; and for fuch care, each fucceffive day would repay in increafed pleafures and enjoyments.

The lungs and the ftomach are too near neighbours not to be mutually affected by abufes offered to the one or the other; they both have their natural food, and the natural and appropriate means prepared by which it is to be received. Air is the efpecial food of the lungs, and not of the Stomach. He who fleeps with his mouth open draws cold air and its impurities into the ftomach as well as into the lungs; and various difeafes of the ftomach, with indigeftion and dyfpepfia, are the confequences. Bread may almoft as well be taken into the lungs, as cold air and wind into the ftomach.

A very great proportion of human difeafes are attributed to the ftomach, and are there met and treated; yet I believe they have a higher origin, the lungs; upon the healthy and regular action of which the digeftive, as well as the refpiratory and nervous fyftems depend; the moving, active, principle of life, and *life itfelf*, are there; and whatever deranges the natural action at that fountain affects every function of the body.

The ftomach performs its indifpenfable, but fecondary part, whilft the moving motive power is in healthy action, and no longer. Man can exift feveral days without food, and but about as many minutes, without the action of his lungs. Men habitually fay " they don't fleep well, becaufe fomething is wrong in their ftomachs," when the truth may be, that their ftomachs are wrong becaufe fomething is wrong in their fleep.

If this dependent affinity in the human fyftem be true, befetting man's life with fo many dangers flowing from the abufe of his lungs, with the fact that the brute creations are exempt from all of thefe dangers, and the favages in the wildernefs nearly fo, how important is the queftion which it raifes whether the frightful and unaccountable Bills of Mortality amongft the civilized Races of mankind are not greatly augmented, if not chiefly caufed, by this error of life, beginning, as I have faid, in the cradle, and becoming by habit, as it were, *a fecond nature*, to weary and torment Mankind to their graves?

Man is created, we are told, to live three fcore and ten years, but how fmall a proportion of mankind reach that age, or half way, or even a quarter of the way to it! We learn from the official Reports before alluded to, that in civilized communities, one half or more perifh in infancy or childhood, and one half of the remainder between that and the age of 25, and Phyficians tell us the difeafes they die of; but who tells us of the *caufes* of thofe difeafes? All effects have their caufes— difeafe is the caufe of death—and there is a caufe for difeafe

When we fee the Brute creations exempted from premature death, and the Savage Races comparatively fo, whilft civilized communities fhow fuch lamentable Bills of Mortality, it is but a rational deduction that that fatality is the refult of habits not practifed by favages and the brute creations; and what other characteriftic differences in the habits of the three creations ftrike us as fo diftinctly different, and fo proportioned to the refults, as already fhown; the *firft*, with the mouth always fhut; the *fecond*, with it fhut during the night and moft of the day; and the *third*, with it open moft of the day and all of the night? The *firft* of thefe are free from difeafe; the *fecond*, comparatively fo; and the *third* fhow the lamentable refults in the Bills of Mortality already given.

How forcible and natural is the deduction from thefe facts, that here may be the great and principal caufe of fuch widely different refults, ftrengthened by the other facts, that the greater part of the fatal difeafes of the body as well as difeafes of the mind, before mentioned, are fuch as could, and *would* flow, from fuch an unnatural abufe of the lungs, the fountain and main-fpring of life; and how important alfo, is the queftion raifed by thefe facts, how far fuch an unnatural habit expofes the human race to the dangers from Epidemic difeafes. The Brute creations are everywhere free from Cholera and Yellow fever, and I am a living witnefs that the Afiatic Cholera of 1831, was everywhere arrefted on the United States frontier, when in its progrefs it reached the Savage tribes living in their primitive condition; having been a traveller on thofe frontiers during its ravages in thofe regions.

Epidemic difeafes are undoubtedly communicated through the medium of the atmofphere, in poifonous animalcules or infectious agents; and what conclusion can be more rational, than that he who sleeps with his mouth open during the night, drawing an increased quantity of infected atmosphere directly on the lungs and into the stomach, will increase his chances of contracting the disease ? And how interesting to Science, and how infinitely important to the *Welfare* of the *Human Race might yet be the enquiry*, whether the thousands and millions of victims to Cholera and yellow fever, were not those very portions of society who were in the habit of sleeping with their mouths open, in the districts infected with those awful scourges!*

It is a well-known fact that fishes will die in a few moments, in their own element, with their mouths kept open by the hook; and I strongly doubt whether a horse or an ox would live any length of time, with its mouth fastened open with a block of wood, during the accustomed hours of its repose; and I believe

---

* My opinions on this important subject having been formed many years ago, as seen in the foregoing pages, I have had opportunities of making observations of an interesting nature, in my recent travels; and amongst those opportunities, one of the most impressive, whilst I was making the voyage on one of the Mail Steamers from Montevideo to Pernambuco, on the coast of Brazil, in the summer of 1857, during which melancholy voyage about 30 out of 80 passengers died of the Yellow fever, and were launched from the deck into the sea, according to the custom. Having been twice tried by that disease on former occasions, and consequently feeling little or no alarm for myself, I gave all my time and attention to the assistance of those who were afflicted. Aware of the difficulty of closing the mouth of a corpse whose mouth has been habitually open through life, and observing that nearly every one launched from the vessel had the character and expression strongly impressive of the results of that habit, I was irresistibly led to a private and secret scanning of faces at the table and on deck, and of six or seven persons for whom I had consequent apprehensions, I observed their seats were in a day or two vacated, and afterwards I recognised their faces, when brought on deck, as subjects for the last, sad ceremony.

that the derangement of the fyftem by fuch an experiment
would be fimilar to that in the human frame, and that death
would be fooner and more certain; and I believe alfo, that if the
American Races of Savages which I have vifited, had treated
this fubject with the fame indifference and abufe, they would
long fince have loft (if not have ceafed to exift) that decided
advantage which they now hold, over the civilized Races, in
manly beauty and fymmetry of phyfical conformation; and that
their Bills of Mortality would exhibit a much nearer approxi-
mation to thofe of civilized communities than they now do.*

Befides the lift of fatal difeafes already given, and which I
attribute chiefly to the pernicious habit which I have explained,
there are other refults affecting the fenfes, perfonal appearance,
and the enjoyments of life, which, though not *fatal*, are themfelves of fufficient importance to demand its correction; fuch
as Curvature of the Spine, Idiocy, Deafnefs, Nightmare, Polypus
in the Nofe, Malformation and premature decay of the teeth,
toothache, tic-douloureux, Rheumatifm, Gout, and many others,
to which the brute creations are ftrangers, and to moft of which
the Savage Races are but little fubject.

By another reference to the Statiftics of civilized Societies,
we find that in *fome*, one half per cent. are Idiots or Lunatics;
one-third per cent. are Deaf and Dumb, one half per cent. are
hunch-backs, and from three-fourths to one per cent. of other
difabling difeafes and deformities; all of which are almoft

---

* I have before faid that the Brute creations are everywhere free from Cholera,
Yellow fever, and other epidemics; yet they are as fubject as the human fpecies,
to the effects of other poifons. Who knows until it is tried, how long a horfe, an
ox, or a dog could exift in one of thofe infected diftricts, with its mouth faftened
open, and its noftrils clofed?

unknown to the American Native Races; affording a ſtrong corroborative proof, if it were neceſſary, that ſuch deficiencies and deformities are the reſults of accidents or habits, and not the works of Nature's hand.

Nature produces no diſeaſes, nor deformities; but the off-ſpring of men and women whoſe ſyſtems are impaired by the habits which have been alluded to, are no doubt oftentimes uſhered into the world with conſtitutional weakneſſes and predi-lections for contracting the ſame habits, with their reſults; and it is ſafe to ſay, that three-fourths of the generating portions of every civilized community exiſting, are more or leſs under theſe diſqualifications, which, together with want of proper care of their offspring, in infancy and childhood, I believe to be the cauſe of four-fifths of the mental and phyſical deformities, loſs of teeth, and premature deaths, between conception and infancy, childhood, manhood, and old age.

I have ſaid that no diſeaſes are natural, and deformities, mental and phyſical, are neither hereditary nor natural, but purely the reſults of accidents or habits. A cloven-foot produces no cloven-feet, hunch-backs beget ſtraight ſpines, and mental defor-mities can have no progeny.

What a ſad bill to bring againſt the glorious advantages of civilized life, its *improvements*, its *comforts* and *refinements*, that in England there are ſomething like 35,000 Idiots and Lunatics—17,000 Deaf and Dumb, and 15,000 hunch-backs, and about an equal proportion of theſe mental and phyſical deformities in the other civilized nations of the Earth!

Nature makes nothing without deſign; and who dares to ſay that ſhe has deſigned theſe liſts of pitiable exiſtence amongſt the civilized Races of Man, and that the more perfect work of her

hand has been beftowed upon the Savage (and even the Brute)
creations ?   And next to Nature, our *dear Mothers*, under whofe
kind care and tender handling we have been raifed, could fubject us to no accident to turn the brain or crook the fpine ; but
eafily and thoughtlefsly might, even in their *over* anxiety for our
health, fubject us to early treatment, engendering habits which
would gradually and imperceptibly produce the whole of thefe
calamities ; which I believe have never, as yet, been traced to a
more probable caufe than the habitual abufe of the lungs, in the
manner which has been defcribed.

The teeth of Man, as with the Brutes, are wifely conftructed
to anfwer their intended purpofes through the natural term of
life, and would fo, no doubt, but from abufes, the principal one
of which I confider to be the pernicious habit already explained.
The faliva exuding from the gums, defigned as the Element of
the teeth, floods every part of the mouth while it is fhut; con-
tinually rifing, like a pure fountain, from the gums, at the roots
of and between the teeth; loofening and carrying off the
extraneous matter which would otherwife accumulate, commu-
nicating difeafe to the teeth, and taint to the breath.

By nature, the teeth and the eyes are ftrictly *amphibious ;* both
immerfed in liquids which are prepared for their nourifhment
and protection, and with powers of exifting in the open air long
enough for the various purpofes for which they were defigned;
but beyond that, abufe begins, and they foon turn to decay.   It
is the fuppreffion of faliva, with drynefs of the mouth, and an
unnatural current of cold air acrofs the teeth and gums during
the hours of fleep, that produces malformation of the teeth,
toothache, and tic-douloureux, with premature decay, and lofs of
teeth fo lamentably prevalent in the civilized world.

Amongſt the Brute creations, that never open their mouths except for taking their food and drink, their teeth are protected from the air both day and night, and ſeldom decay; but with Man, who is a *talking* and *laughing* animal, expoſing his teeth to the air a great portion of the day, and oftentimes during the whole of the night, the reſults are widely different—he is often-times toothleſs at middle age, and in ſeven caſes in ten, in his grave before he is fifty.

If civilized man, with his uſual derangements and abſence of teeth, had been compelled to crop the graſs, like the ox and the horſe, as the means of his living, and knew not the glorious uſe of the *Spoon*, to what a miſery would he have been doomed, and how long could he exiſt? the loſs of a tooth or two with thoſe animals would reſult in their death; and how wiſe, and how provident therefore, the deſigns of the Creator, who has provided them with the unfailing means of ſupporting their exiſtence, and alſo the inſtinctive habits intended for the *protection* of thoſe means.

Amongſt the Native Races they ſeem to have a knowledge of theſe facts; and the poor Indian woman who watches her infant and preſſes its lips together as it ſleeps in its cradle attracts the ridicule perhaps, or pity, of the paſſer-by, but ſecures the habit in her progeny which enables them to com-mand the admiration and envy of the world.

Theſe people, who talk little and ſleep naturally, have no Dentiſts nor dentifrice, nor do they require either; their teeth almoſt invariably riſe from the gums and arrange themſelves as regular as the keys of a piano; and without decay or aches, preſerve their ſoundneſs and enamel, and powers of maſtication, to old age: and there are no ſufficient reaſons aſſigned yet, why the fame refults, or nearly fuch, may not be produced amongſt the more enlightened Races, by ſimilar means.

Civilized man may properly be ſaid to be an *open mouthed animal;* a wild man is not. An Indian Warrior ſleeps, and hunts, and ſmiles, with his mouth ſhut; and with ſeeming reluctance, opens it even to eat or to ſpeak. An Indian child is not allowed to ſleep with its mouth open, from the very firſt ſleep of its exiſtence; the conſequence of which is, that while the teeth are forming and making their firſt appearance, they *meet* (and conſtantly *feel*) each other; and taking their relative, natural poſitions, form that healthful and pleaſing regularity which has ſecured to the American Indians, as a Race, perhaps the moſt manly and beautiful mouths in the World.*

---

* When I ſpeak of comparative perſonal appearance or of the habits of a people, I ſpeak of them collectively, and in the aggregate. I often ſee mouths and other phyſical conformations amongſt the civilized portions of mankind equally beautiful as can be ſeen amongſt the Savage Races, but by no means ſo often. Symmetry of form, gracefulneſs of movement, and other conſtituents of manly beauty are much more *general* amongſt the Savage Races; and their Societies, free from the humbled and dependent miſery which *comparative* poverty produces in Civilized Communities, produce none of thoſe ſtriking contraſts which ſtare us in the face, and excite our diſguſt and our ſympathies, at nearly every ſtep we take. The American Savages are all poor, their higheſt want is that of food, which is generally within their reach; their faces are therefore not wrinkled and furrowed with the ſtamp of care and diſtreſs, which extreme poverty begets—the repulſive marks which avarice engraves, nor with the loathſome and diſguſting expreſſions which the prodigal diſſipations of Wealth often engender in Civilized Societies. Their taſtes and their paſſions are leſs refined and leſs ardent, and more ſeldom exerted, and conſequently leſs abuſed; they live on the ſimples of life, and imagine and deſire only in proportion; the conſequences of which are, that their faces exhibit ſlighter inroads upon Nature, and conſequently a greater average of good looks than an equal community of any civilized people.

Nature makes no derangements or deformities in teeth or mouths; but habits or accidents produce the difagreeable derangements of the one, and confequently the difgufting expreffions of the other, which are fo often feen.

With the brute creations, where there is lefs chance for habits or accidents to make derangements, we fee the beautiful *fyftem* of the regularity of the works of Nature's hand, and in their foundnefs and durability, the *completenefs* of her works, which we have no juft caufe to believe has been ftinted in the phyfical conftruction of man.

The contraft between the two Societies, of Savage and of Civil, as regards the perfection and duration of their teeth, is quite equal to that of their Bills of Mortality, already fhown; and I contend that in both cafes, the principal caufe of the difference is exactly the fame, that of refpiration through the mouth, during the hours of fleep.

Under the lefs cruel, and apparently more tender and affectionate treatment, of many civilized mothers, their infants fleep in their arms, in their heated exhalation, or in cradles, in overheated rooms, with their faces covered, without the allowance of a breath of vital air; where, as has been faid, they from neceffity gafp for breath until it becomes a habit of their infancy and childhood, to fleep with their mouths wide open, which their tender mothers overlook, or are not cruel enough to correct; little thinking of the fad affliction which the croup, or later difeafes are to bring into their houfe.

There is nothing more natural than a mother's near and fond embrace of her infant in her hours of fleep; and nothing more dangerous to its health, and even to its exiftence. The tender fympathies of love and inftinct draw her arms clofer around it and her lips nearer, as fhe finks into fleep and compels it to breathe the exhaufted and poifoned air that fhe exhales from her own lungs; little thinking how much fhe is doing to break her heart in future days. Nothing is fweeter or more harmlefs to a mother than to inhale the feeble breath of her innocent; but fhe fhould be reminded that whilft fhe is *drawing* thefe delicious draughts, fhe may be *returning* for them, peftilence and death.

All mothers know the painful, and even dangerous crifis which their infants pafs in teething; and how naturally do their bofoms yearn for the fufferings of thefe little creatures whofe Earthly careers are often ftopped by that event. (3,660 per annum in England alone, under one year of age, as has been fhown.)

Amongft the Savage Races, we have feen that death feldom, if ever, enfues from this caufe; and how eafy it is to perceive that unnatural pains, and even death, may be caufed by the habit of infants fleeping with their mouths ftrained open, and expofed to the cold air, when the germs of the teeth are firft making their appearance.

The Statiftics of England fhow an annual return of "25,000 infants, and children under five years of age, that die of *convulfions.*" What caufes fo probable for thofe convulfions as teething and the croup; and what more probable caufe for the *unnatural* pains of teething and the croup, than the *infernal* habit which I am condemning.

At this tender age, and under the kind treatment juft mentioned, is thoughtleffly laid the foundation for the rich harvefts which the Dentifts are reaping in moft parts of the civilized world. The infant paffes two-thirds of its time in fleep, with

its mouth open, while the teeth are prefenting themfelves in
their tender ftate, to be chilled and dried in the currents of air
paffing over them, inftead of being nurtured by the warmth and
faliva intended for their protection, when they project to unnatural
and unequal lengths, or take different and unnatural directions,
producing thofe difagreeable and unfortunate combinations,

which are frequently feen in civilized adult focieties, and oftentimes fadly disfiguring the human face for life.

While there are a great many perfons in all civilized focieties who adhere to the defigns of Nature in the habits above
referred to, how great a proportion of the individuals of thofe
focieties carry on their faces the proofs of a different habit,
brought from their childhood, which their Conftitutions have fo
far fuccefsfully battled againft, until (as has been faid) it becomes
like a fecond Nature, and a matter of *neceffity*, even during
their waking hours and the ufual avocations of life, to breathe
through the mouth, which is conftantly open; while the
nafal ducts, being vacated, like vacated roads that grow up

to grafs and weeds, become the feat of Polypus and other difeafes.

In all of thefe inftances there is a derangement and deformity of the teeth, and disfigurement of the mouth, and the *whole face*, which are not natural; carrying the proof of a long practice of the baneful habit, with its lafting confequences; and producing that unfortunate and pitiable, and oftentimes difgufting expreffion, which none but civilized communities can prefent.

Even the Brute creations furnifh nothing fo abominable as thefe; which juftly demand our *fympathy* inftead of our *derifion*. The faces and the mouths of the Wolf, the Tiger, and even the Hyena and the Donkey, are agreeable, and even handfome, by the fide of them.

What Phyfician will fay that the inhalation of cold air to the lungs through fuch mouths as thefe, and over the putrid fecretions and rotten teeth within, may not occafion difeafe of the lungs and death? Infected diftricts communicate difeafe—

infection attaches to putrefcence, and no other infected diftrict can be fo near to the lungs as an infected mouth.

Moft habits againft Nature, if not arrefted, run into difeafe. The habit which has thus far been treated as a *habit*, merely, with its evil confequences, will here be feen to be worthy of a *name*, and of being ranked amongft the fpecific difeafes of mankind ; Indulged and practifed until the mouth is permanently diftorted from its natural fhape, and in the infectious ftate above named, acting the unnatural hand-maid to the lungs, it gains the locality and fpeciality of character which characterize difeafes, and therefore would properly rank amongft them. No name feems as yet to have been applied to this malady, and no one apparently more expreffive, at prefent fuggefts, than *Malo inferno*, which (though perhaps not exactly Claffic) I would *denominate* it, and *define* it to be ftrictly a *human* difeafe, confined chiefly to the Civilized Races of Man, an unnatural and pitiable disfigurement of the " human face divine," *unknown* to the Brutes, and *unallowed* by the Savage Races, *caufed* by the carelefs permiffion of a habit contracted in infancy or childhood, and fubmitted to, humbly, through life, under the miftaken belief that it is by an unfortunate order of Nature—its *Remedy* (in neglect of the fpecifics to be propofed in the following pages) the *grave* (generally) between infancy and the age of forty.

The American Indians call the civilized Races "*pale faces*" and "*black mouths*," and to underftand the full force of thefe expreffions, it is neceffary to live awhile amongft the Savage Races, and then to return to civilized life. The Author has had ample opportunities of tefting the juftnefs of thefe expreffions, and has been forcibly ftruck with the correctnefs of their application, on returning from Savage to Civilized Society. A long

familiarity with red faces and clofed mouths affords a new view
of our friends when we get back, and fully explains to us the
horror which a favage has of a " pale-face," and his difguft with
the expreffion of open and *black mouths*.*

No man or woman with a handfome fet of teeth keeps the
mouth habitually open; and every perfon with an unnatural
derangement of the teeth is as fure feldom to have it fhut.
This is not becaufe the derangement of the teeth has made the
habit, but becaufe the habit has caufed the derangement of the
teeth.

If it were for the fake of the teeth alone, and man's perfonal appearance, the habit I am condemning would be one
well worth ftruggling againft; but when we can fo eafily, and
with fo much certainty difcover its deftructive effects upon the
conftitution and *life* of man, it becomes a fubject of a different
importance, and well worthy of being underftood by every
member of fociety, who themfelves, and not phyficians, are to
arreft its deadly effects.

The Brute, at its birth, rifes on its feet, breathes the open
air, and feeks and obtains its food at the next moment.

---

* Of the party of 14 Ioway Indians, who vifited London fome years fince,
there was one whofe name was Wafh-ke-mon-ye (the faft dancer) ; he was a great
droll, and fomewhat of a critic ; and had picked up enough of Englifh to enable
him to make a few fimple fentences and to draw amufing comparifons. I afked
him one day, how he liked the White people, after the experience he had now
had ; to which he replied—" Well, White man—fuppofe—mouth fhut, putty coot,
mouth open, no coot—me no like um, not much." This reply created a fmile
amongft the party, and the chief informed me that one of the moft ftriking pecu-
liarities which all Indian Tribes difcovered amongft the white people, was the
derangement and abfence of their teeth, and which they believed were deftroyed
by the number of ies that paffed over them.

The Chicken breaks its own fhell and walks out on two legs, and without a gaze of wonder upon the world around, begins felecting and picking up its own food!

Man, at his birth, is a more helplefs animal, and his mental, as well as his phyfical faculties, requiring a much longer time to mature, are fubject to greater dangers of mifdirection from pernicious habits, which it fhould be the firft object of parents to guard againft.

The Savage Tribes of America allow no obftacles to the progrefs of Nature in the development of their teeth and their lungs for the purpofes of life, and confequently fecuring their exemption from many of the pangs and pains which the civilized Races feem to be heirs to; who undoubtedly too often *over*-educate the *intellect*, while they *under*-educate the *Man*.

The human infant, like the infant brute, is able to breathe the natural air at its birth, both afleep and awake; but that breathing fhould be done as Nature defigned it, through the noftrils, inftead of through the mouth.

The Savage Mother, inftead of embracing her infant in her fleeping hours, in the heated exhalation of her body, places it at her arm's length from her, and compels it to breathe the frefh air, the coldnefs of which generally prompts it to fhut the mouth, in default of which, she preffes its lips together in the manner that has been ftated, until fhe fixes the habit which is to laft it through life; and the contraft to this, which is too often practifed by mothers in the civilized world, in the miftaken belief that *warmth* is the effential thing for their darling babes, I believe to be the innocent foundation of the principal, and as yet unexplained caufe of the deadly difeafes fo frightfully fwelling the Bills of Mortality in civilized communities.

All Savage infants amongſt the various native Tribes of
America, are reared in cribs (or cradles) with the back laſhed

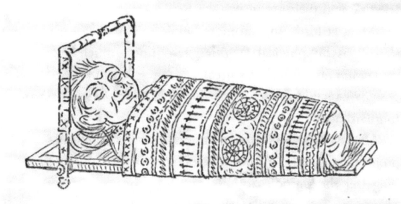

to a ſtraight board; and by the aid of a circular, concave
cuſhion placed under the head, the head is bowed a little forward when they ſleep, which prevents the mouth from falling
open; thus eſtabliſhing the early habit of breathing through
the noſtrils. The reſults of this habit are, that Indian adults
invariably walk erect and ſtraight, have healthy ſpines, and ſleep

upon their backs, with their Robes wrapped around them; with
the head ſupported by ſome reſt, which inclines it a little forward, or upon their faces, with the forehead reſting on the arms

which are folded underneath it, in both of which cafes there is a tendency to the clofing of the mouth; and their fleep is therefore always unattended with the nightmare or fnoring.

Lying on the back is thought by many to be an unhealthy practice; and a long habit of fleeping in a different pofition may even make it fo; but the general cuftom of the Savage races, of fleeping in this pofition from infancy to old age, affords very conclufive proof, that if commenced in early life, it is the healthieft for a general pofture, that can be adopted.

It is very evident that the back of the head fhould never be allowed, in fleep, to fall to a level with the fpine; but fhould be fupported by a fmall pillow, to elevate it a little, without raifing the fhoulders or bending the back, which fhould always be kept ftraight.

The Savages with their pillows, like the birds in the building of their nefts, make no improvements during the lapfe of ages, and feem to care little if they are blocks of wood or of ftone, provided they elevate the head to the required pofition.

With the civilized Races, where everything is progreffive, and luxuries efpecially fo, pillows have increafed in longitudinal dimenfions until they too often form a fupport for the fhoulders as well as the head, thereby annulling the object for which they were originally intended, and for which, alone, they should be ufed.

All animals lower the head in fleep; and mankind, with a fmall fupport under it, inclining it a little forward, affume for it a fimilar pofition.

This elderly and excellent Gentleman, from a long (and

therefore neceffary) habit, takes his nap after dinner, in the attitude which he is contented to believe is the moft luxurious that can be devifed; whilft any one can difcover that he is very far from the actual enjoyment which he might feel, and the more agreeablenefs of afpect which he might prefent to his furrounding friends, if his invention had carried him a little farther, and fuggefted the introduction of a fmall cufhion behind his head, advancing it a little forward, above the level of his fpine. The gaftric juices commence their work upon the frefh contents of a ftomach, on the arrival of a good dinner, with a much flighter jar upon the digeſtive and nervous ſyſtems, when the ſoothing and deleɥable compound is not ſhocked by the unwelcome inhalations of chilling atmoſphere.

And this tender and affeɥionate Mother, *bleſſing* herſelf and her flock of little ones with the *pleaſures of ſleep!* how much

might ſhe increaſe her own enjoyment with her pillow under her *head*, inſtead of having it under her *ſhoulders;* and that of her little gaſping innocents, if ſhe had placed them in cribs, and with pillows under their heads, from which they could not eſcape.

The contraſt between the expreſſions of theſe two groups

will be ftriking to all; and every mother may find a leffon in
them worth her ftudying; either for improvements in her own
Nurfery, or for teaching thofe who may ftand more in need of
Nurfery Reform than herfelf.

So far back as the ftarting point in life, I believe man feldom
looks for the caufes of the pangs and pains which befet and
torture him in advanced life; but in which, far back as it may
be, they may have had their origin.

Little does he think that his aching, deformed, and decaying teeth were tortured out of their natural arrangement and
health, in the days of their formation, by the cold draughts of
air acrofs them; or that the confumption of his decaying lungs
has been caufed by the fame habit; and that habit was the
refult of the actual tendernefs, but overfight, of his affectionate
Mother, when he flept in her arms, or in the cradle.

The foregoing are general remarks which I have been
enabled to make, from long and careful obfervation; and
there are others perhaps, equally or more demonftrative of the
danger of the habit alluded to, as well as of the power we have
of averting it, and of arrefting its baneful effects, even in middle
age, or the latter part of man's life, which will be found in the
relation of my own experience.

At the age of 34 years, (after devoting myfelf to the dry
and tedious ftudy of the Law for 3 years, and to the practice of
it for 3 years more, and after that to the ftill more fatiguing and
confining practice of Miniature and portrait painting, for 8
years,) I penetrated the vaft wildernefs with my canvafs and
brufhes, for the purpofe which has already been explained; and
in the profecution of which defign. I have devoted moft of the
fubfequent part of my life.

At that period I was exceedingly feeble, which I attributed to the sedentary habits of my occupation, but which many of my friends and my Phyſician believed to be the reſult of diſeaſe of the lungs. I had, however, no apprehenſions that dampened in the leaſt, the ardour and confidence with which I entered upon my new ambition, which I purſued with enthuſiaſm and unalloyed ſatisfaction until my reſearches brought me into ſolitudes ſo remote that beds, and bed chambers with fixed air, became matters of impoſſibility, and I was brought to the abſolute neceſſity of ſleeping in canoes or hammocks, or upon the banks of the Rivers, between a couple of Buffalo ſkins, ſpread upon the graſs, and breathing the chilly air of dewy and foggy nights, that was circulating around me.

Then commenced a ſtruggle of no ordinary kind, between the fixed determination I had made, to accompliſh my new ambition, and the daily and hourly pains I was ſuffering, and the diſcouraging weakneſs daily increaſing on me, and threatening my ultimate defeat.

I had been, like too many of the world, too tenderly careſſed in my infancy and childhood, by the over kindneſs of an affectionate Mother, without cruelty or thoughtfulneſs enough to compel me to cloſe my mouth in my ſleeping hours; and who, through my boyhood, thinking that while I was aſleep I was doing well enough, allowed me to grow up under that abominable cuſtom of ſleeping, much of the time, with the mouth wide open; and which practice I thoughtleſſly carried into manhood, with Nightmare and ſnoring, and its other reſults; and at laſt, (as I diſcovered juſt in time to ſave my life,) to the banks of the Miſſouri, where I was nightly drawing

4

the deadly draughts of cold air, with all its poifonous malaria,
through my mouth into my lungs.

Waking many times during the night, and finding myfelf
in this painful condition, and fuffering during the fucceeding
day with pain and inflammation (and fometimes bleeding) of
the lungs, I became fully convinced of the danger of the habit,
and refolved to overcome it, which I eventually did, only by
fternnefs of refolution and perfeverance, determining through
the day, to keep my teeth and my lips firmly clofed, except
when it was *neceffary* to open them; and ftrengthening this
determination, as a *matter* of *life or death*, at the laft moment of
confcioufnefs, while entering into fleep.

Under this unyielding determination, and the evident relief
I began to feel from a partial correction of the habit, I was
encouraged to continue in the unrelaxed application of my
remedy, until I at length completely conquered an infidious
enemy that was nightly attacking me in my helplefs pofition,
and evidently faft hurrying me to the grave.

Convinced of the danger I had averted by my own perfeverance, and gaining ftrength for the continuance of my daily
fatigues, I renewed my determinations to enjoy my natural
refpiration during my hours of fleep, which I afterwards did,
without difficulty, in all latitudes, in the open air, during my
fubfequent years of expofure in the wildernefs; and have fince
done fo to the prefent time of my life; when I find myfelf
ftronger, and freer from aches and pains than I was from my
boyhood to middle age, and in all refpects enjoying better health
than I did during that period.

I mention thefe facts for the benefit of my fellow beings, of
whom there are tens (and hundreds) of thoufands fuffering

from day to day from the ravages of this infidious enemy that preys upon their lungs in their unconfcious moments, who know not the caufe of their fufferings, and find not the Phyfician who can cure them.

Finding myfelf fo evidently relieved from the painful and alarming refults of a habit which I recollected to have been brought from my boyhood, I became forcibly ftruck with the cuftom I had often obferved (and to which I have before alluded) of the Indian women preffing together the lips of their fleeping infants, for which I could not, at firft, imagine the motive, but which was now fuggefted to me in a manner which I could not mifunderftand; and appealing to them for the object of fo, apparently, cruel a mode, I was foon made to underftand, both by their women and their Medicine Men, that it was done " to enfure their good looks, and prolong their lives;" and by looking into their communities, and contrafting their fanitary condition with the Bills of Mortality amongft the civilized Races, I am ready to admit the juftnefs of their reply; and am fully convinced of the advantages thofe ignorant Races have over us in this refpect, not from being *ahead* of us, but from being *behind* us, and confequently not fo far departed from Nature's wife and provident regulations, as to lofe the benefit of them.

From the whole amount of obfervations I have made amongft the two claffes of fociety, added to my own experience, as explained in the foregoing pages, I am compelled to believe, and feel authorifed to affert, that a great proportion of the difeafes prematurely fatal to human life, as well as mental and phyfical deformities, and deftruction of the teeth, are caufed by the abufe of the lungs, in the Mal-refpiration of Sleep: and alfo, that the pernicious habit, though contracted in infancy or childhood, or manhood, may generally be corrected by a fteady and determined perfeverance, bafed upon a conviction of its baneful and fatal refults.

The great error is moft frequently committed, and there is the proper place to correct or prevent it, at the *ftarting point*— when the germs are tender, and taking their firft impreffions, which are to laft them through life. It is then too, that the fondeft and tendereft fympathies belonging to the human breaft are watching over them; and it is only neceffary for thofe kind guardians to be made aware of the danger of thoughtlefs habits which their over-indulgence may allow their offspring to fall into.

It is to *Mothers*, and truly not to Phyficians or Medicines, that the world are to look, for the remedy of this evil; and the phyfical improvements of mankind, and the prolongation of human exiftence, effected by it.

Children, I have faid, are not born *Hunch-backs*, but a habit of fleeping thus, in the varying temperatures of the night,

might make them fuch. Infants are not born *Idiots* or *Lunatics*,

but a habit of sleeping thus, in sudden changes of weather,

would tend to make them so, and in the countries where infants sleep thus, the above deformities scarcely exist; while in Eng-

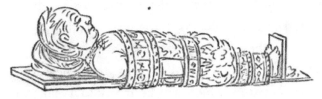

land, as has been shown, there are 20,000 of the first of these, and 35,000 of the latter. How significant and important the deductions from these simple facts—if they be facts—and who will contradict them?

If Physicians and Surgeons gain fame for occasionally conquering the enemy in *combat;* what laurels, and what *new Title,* should await the *fair Diplomatists* who will *keep the enemy out of the field*—the affectionate *mothers,* who, like the Indian woman, will sit by their sleeping infants, and watch and guard them through their childhood, against the departure from one of Nature's most wise and important regulations, designed for their health and happiness.

If the great majority of this sort of evil has its origin in that early period of life, its correction comes directly under the Mother's province; and there certainly can be no better gua-

rantee for the benefit of coming generations, than that mothers
fhould be made fully fenfible of the evil, and of their own
power to avert it. And TO *Mothers*, I would in the firft place,
say, for the fakes of your infants unborn, and for your own
lives' sake, draw the curtain, (not of your bed, but of your
lungs) when you retire to reft; availing yourfelves and your
offspring of the full benefit of the peaceful and invigorating
repofe which Nature has prepared for you, to enable you to
meet with fuccefs the events to which you are approaching;
and when Nature has placed in your arms for your kind care,
the darling objeȼts of your tendereft affeȼtions, not to forget that
fhe has prepared and defigned them to breathe the open air;
and that when they fleep in your embrace in heated rooms and
feather beds, they fleep in a double or treble heat, the thoughtlefs confequences of which will be likely to break your hearts
in future life. Reft affured that the great fecret of life is the
breathing principle, for which Nature has rightly prepared the
material, and the *proper mode of ufing it;* and at the incipient ftage
of life where *Mothers* are the Phyficians, is the eafieft place to
contraȼt habits againft Nature, or to correȼt them; and that there
is *woman's* poft, her appropriate fphere; where fhe takes to herfelf the fweeteft pleafures of her exiftence, and draws the higheft
admiration of the World, whilft, like a *guardian Angel*, fhe is
watching over, and giving direȼtion to, the Deftinies of Man.

To Children—to Boys and Girls, who have grown up to the
age of difcretion, and are able to read, the above information
and advice are doubly important, becaufe you have long lives
of enjoyment or mifery before you; and which, you now being
out of your Mother's immediate care, are to be controlled by

your own actions. And that you may not undervalue the advice which I am about to advance directly to you, I may (as the Clergyman repeats his text in his Sermon, or a fond parent, the important points of his advice to his fon) repeat fome things that I have faid, while I am giving you *further* evidence of the importance of the fubject I am now explaining to you.

I advife you to bear in mind the awful Bills of Mortality amongft civilized focieties, which I have quoted; and realize the dangerous race which civilized man runs in life—how very few live to the age defigned by Nature—how many perifh in infancy, long before they are of your age; and confequently the dangers which you have already paffed, and contraft all of thefe with thofe of the Wild Indians, who by Nature, are no ftronger than we are, but who generally live to good old age, with comparatively, few bodily pains in life, and their teeth almoft uniformly regular and found, without the aid of Dentifts and tooth-brfhues.

Have you obferved by thofe Bills of Mortality, that you are but one out of two or three of your little companions who ftarted and commenced playing along with you, permitted to live to boyhood; and alfo that you have but one chance in four, or thereabouts, of living to tolerable old age?

Can you read thofe lamentable eftimates, which are matters of fact, and draw fuch fearful conclufions from them as to your own condition and profpects, without realizing the importance of the fubject? and can you compare thofe difafters amongft the civiljzed, with thofe of the Savage Races, which I have explained, without believing there is fome caufe for all this, that is unnatural, and which may be, to a great degree, corrected, if we make the proper effort?

You have read in the foregoing pages, that man's life depends from one moment to another on the air which he breathes, and alfo that the atmofphere is nowhere pure enough for the healthy ufe of the lungs until it has paffed the purifying procefs which Nature has prepared in the noftrils, and which has been explained. Air is an Elementary principle, created by the hand of God, who, as has been faid, creates nothing but perfections; and confequently is nowhere impure, except from the caufes which I have already explained; and in the infinity of His wifdom and goodnefs, thofe accidental impurities were forefeen and provided for (even with the brutes, as well as with Mankind), by the myfterious organizations through which the breath of life firft came to man.

The various occupations of men, and for which you are by this time preparing, fubject them more or lefs to the dangerous effects of the malaria and poifonous particles in the air, in proportion to the nature of their employments, and the diftricts and atmofpheres in which they exift and work.

The Mechanical trades are the moft fubject to thefe, from which the farmer and the Gentleman are more exempt; the Carpenter, therefore, amidft the duft of his fhop, fhould work with his mouth fhut, and take care not to fleep upon his bench during his mid-day reft. The Cutlery grinder fhould not work with his mouth open amidft the particles of fteel which his feet raife from the floor, and the motion of his wheel keeps in circulation in the air.

So with the Stone-cutter (and particularly thofe working in the hardeft fort of ftones and flint) the fame precautions are neceffary; as by the extraordinary proportion of deaths Reported amongft thofe claffes of workmen, the poifonous effects of their bufinefs are clearly proved, as well as by the accumulated particles of fteel and filex found imbedded in their lungs and coating the Refpiratory organs; and which, to have caufed premature death, muft have been inhaled through the *mouth*. Phyficians are conftantly informing the world, in their Reports, of the fatal refults of thefe poifonous things inhaled into the lungs; but why do they not fay at the fame time, that there are two modes of inhalation, by the *nofe* and by the *mouth;* and inform the Mechanics and labourers of the World who are thus rifking their lives, that there is fafety to life in one way, and great danger in the other? If Phyficians forget to give you this advice, thefe fuggeftions, with your own difcretion, may be of fervice to you.

The Savages have the advantage of moving about, and fleeping in the open air; and Civilized Races have the advantages over the poor Indians, of comfortable houfes and beds, and bed-rooms; and alfo of the moft fkilful Phyficians, and Surgeons, and Dentifts; and ftill we are ftruck with the deplorable refults in our fociety, of fome latent caufe of difeafes, which I believe has been too much overlooked and neglected.

Have you not many times waked in the middle of the night, in great diftrefs, with your mouths wide open, and fo cold and dry that it took you a long time to moiften and fhut them again? and did it occur to you at thofe moments that this was all the refult of a carelefs habit, by which you were drawing an unnatural draught of cold air in every breath, directly on the lungs, inftead of drawing it through the noftrils, which Nature has made for that efpecial purpofe, giving it warmth, and meafuring its quantity, fuitable to the demands of repofe?

Watch your little Brothers and Sifters, or other little innocent playfellows, when afleep with their mouths ftrained open,
and obferve the painful expreffions of their faces—their ner-

vous agitation—the unnatural beating of their hearts—the
twitching of their flefh, and the cords of their necks and throats;
and your own reafon will tell you that they do not enjoy fuch
fleep.   And on the other hand, what pictures of innocence and
enjoyment are thofe who are quietly fleeping with their mouths
firmly fhut, and their teeth clofed, fmiling as they are enjoying

their natural repofe? If you will for a few moments fhut
your eyes, and let your under jaw fall down, as it fometimes
does in your fleep, you will foon fee how painful the over

draught of cold air on the lungs becomes, even in the day-time, when all your energies are in action to relieve you; and you will inftantly perceive the mifchief that fuch a mode of breathing might do in the·night, when every mufcle and nerve in your body is relaxed and feeking repofe, and the chill of the midnight air is increafing.

It is, moft undoubtedly, the above named habit which produces *confirmed Snorers,* and alfo confumption of the lungs and many other difeafes, as well as premature decay of the teeth— the Nightmare, &c., from which it has been fhown, the Savage Races are chiefly exempt; and (I firmly believe) from the fact that they always fleep with their mouths clofed, and their teeth together, as I have before defcribed.

There are many of you who read, to whom this advice will not be neceffary, while many others of your little companions will attract your fympathy when you fee them afleep, with their mouths ftrained open, and their fenfations anything but thofe of joy and reft. Their teeth are growing during thofe hours,

and will grow of unequal lengths, and in unnatural directions, and oftentimes difabling them in after life, from fhutting their

mouths, even in their waking hours, and moft lamentably disfiguring their faces for the remainder of their days.

It is then, my young Readers, for you to evade thefe evils,
to fave your own lives and your good looks, by *your own* efforts,
which I believe the moft of you can do, without the aid of Phyficians or Dentifts, who are always the ready and bold antagonifts
of difeafe, but never called until the enemy has made the attack.

I imagine you now juft entering upon the ftage of life,
where you are to come under the gaze of the world, and to
make thofe impreffions, and form thofe connexions in fociety
which are to attend you, and to benefit or to injure you through
life.   You are juft at that period of your exiftence when the
proverb begins to apply, that " man's life is in his own hands ;"
and if this be not always true, it is *quite true*, that much of his
good looks, his daily enjoyments, and the control of his habits,
are within the reach of his attainment.   Thefe are all advantages
worth ftriving for, and if you fternly perfevere for their accom-
plifhment, you will perfectly verify in your own cafes, the other and
truer adage, that " at middle age, man is his own beft Phyfician."

I recollect, and never fhall forget while I live, that in my
boyhood, I fell in love with a charming little girl, merely becaufe
her pretty mouth was always fhut ; her words, which were few,
and always (I thought) fo fitly fpoken, feemed to iffue from
the centre of her cherry lips, whilft the corners of her mouth
feemed (to me,) to be honeyed together.   No excitements
could bring more than a fweet fmile on her lips, which feemed
to hold confident guard over the white and pretty treafures they
enclofed, and which were permitted but occafionally, to be feen
peeping out.

Of fuch a mouth it was eafy to imagine, even without feeing

them, the beautiful embellifhments that were within, as well as the fweet and innocent expreffion of its repofe, during the hours of fleep; and from fuch impreffions, I recollect it was exceedingly difficult and painful to wean my boyifh affections.

To young people, who have the world before them to choofe in, and to be chofen; next to the importance of life itfelf, and their *Future* welfare, are the habits which are to disfigure and impair, or to beautify and protect that feature which, with man and with woman, alike, is the moft expreffive and attractive of the face; and at the fame time, the moft fubject to the influence of pleafing, or difagreeable, or difgufting habits.

Good looks and other perfonal attractions are defirable, and *licenfed* to all; and much more generally attainable than the world fuppofe, who take the various features and expreffions which they fee in the multitude, as the works of Nature's hand.

The natural mouth of man is always an expreffive and agreeable feature; but the departures from it, which are caufed by the predominance of different paffions or taftes, or by the perfectly infipid and difgufting habit which has been explained, are anything but agreeable, and but little in harmony with the advance of his intellect.

Open mouths during the night are fure to produce open mouths during the day; the teeth protrude, if the habit be commenced in infancy, fo that the mouth can't be fhut, the natural expreffion is loft, the voice is affected, polypus takes poffeffion of the nofe, the teeth decay, tainted breath enfues, and the lungs are deftroyed. The whole features of the face are changed, the under jaw, unhinged, falls and retires, the cheeks are hollowed, and the cheek-bones and the upper jaw advance, and the

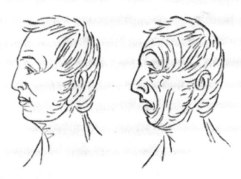

Nature   changed by habit.

brow and the upper eyelids are unnaturally lifted; prefenting at once, the leading features and expreffion of *Idiocy*.

These are changes in the contour and expreffion of the face which any one can fufficiently illuftrate, with a little effort, on his own face before a looking-glafs; and that thefe refults are often fixed and permanently retained in fociety, every fane perfon is able to difcover; and I believe moft perfons will agree with me, that they are the unfortunate refults of the habit I am denouncing.

All the World judge of men's difpofitions and charader by the expreffions of their face; and how difaftrous may it therefore be for men to indulge an expreffion of face in their fleep which they would be afhamed of in their waking hours? The world is full of fuch, however, and fuch a man afleep, and a fleeping Idiot, are exactly the fame.

How appalling the thought, and dangerous the habit! and what are likely to be the refults fhown in the fixed and lafting expreffions of the face?

Thefe remarks, and thefe queftions are intended for *Boys* and *Young Men*, for I can fcarcely allow myfelf to believe that *Young Ladies* would be caught fleeping thus; but one word of advice, even to *them*, may not be amifs—*Idiots afleep* cannot be *Angels awake.*

The natural mouths of mankind, like thofe of the brutes, have a general fyftematic form and expreffion; but the various habits and accidents of life give them a vaft variety of expreffions; and the greater portion of thofe deviations from Nature, are caufed by the malformation of the teeth, or by the falling

of the under jaw, which alone, in its intended pofition, forms the
natural mouth. When formed in this way, and unchanged by
habit or accident, the mouth is always well-fhaped and agreeable;
but if the teeth become deranged in the manner I have
defcribed, the mouth becomes deformed; and in endeavouring
to hide that deformity, oftentimes more difagreeable and unnatural than when that deformity is expofed.

I knew a young Lady many years ago, amiable and intelli-
gent; and agreeable in everything excepting the unfortunate
derangement and fhapes of her teeth; the front ones of which,
in the upper jaw, protruding half an inch or more forward of
the lower ones, and quite incapable of being covered by the
lip, for which there was a conftant effort; the refult of which
was a moft pitiable expreffion of the mouth, and confequently
of the whole face, with continual embarraffment and unhappi-
nefs of the young Lady, and fympathy of her friends. With
all the other charms requifite to have foothed and comforted
the life of any man, fhe lived a life of comparative folitude;
and a few years fince, after a lapfe of 30 years, I met her again;
and though in her old age, fhe was handfome,—her teeth were
all gone, and her lips, from the natural fweetnefs and ferenity
of her temper, feemed to have returned to their native and
childifh expreffion, as if making up for the unnatural and painful fervitude they had undergone.

The human mouth, with the great variety of duties it has
to perform, is fubject to a fufficient variety of expreffions and
diftortions from abufe, independent of thofe arifing from the
habit I am condemning.

The Ear, the Nofe, and the Eyes, being lefs mutable, and
lefs liable to change of character and fhapes, feldom lofe their

natural expreffion; while original nature is as feldom feen remaining in the expreffion of the adult mouth.

This feature, from the variety of its powers and ufes, as well as expreffions, is undoubtedly the greateft myftery in the *material* organization of man. In infant Nature it is always innocent and fweet, and fometimes is even fo in adult life.

Its endlefs modulations of found may produce the richeft, the fweeteft of mufic, or the moft frightful and unpleafant founds in the world. It converfes, it curfes, and applauds; it commends and reproves, it flanders, it flatters, it prays and it profanes, it blafphemes and adores—blows hot and blows cold—fpeaks foft tones of love and affection, and rough notes of vengeance and hatred; it bites, and it woos—it kiffes, ejects faliva, eats cherries, Roaft Beef, and Chicken, and a thoufand other things—drinks coffee, gin, and Mint-juleps (and fometimes Brandy), takes pills, and Rhubarb and Magnefia—tells tales, and keeps fecrets, is pretty, or is ugly, of all fhapes, and of all fizes, with teeth white, teeth black, and teeth yellow, and with no teeth at all.

During the *day*, it is generally eating, drinking, finging, laughing, grinning, pouting, talking, fmoking, fcolding, whiftling, chewing, or fpitting, all of which have a tendency to keep it open; and if allowed to be open during the *night*, is feen, as has been defcribed, by its derangement of the teeth, to create thereby, its own worft deformity.

How ftrange is the fact, that of the three creations—the Brute, the Savage, and the Civilized Races—the ftupid and irrational are taught to perfectly protect and preferve their teeth, through the natural term of life; the ignorant, Savage Races of mankind, with judgment enough *comparatively* to do fo; when enlightened man, with the greateſt amount of knowledge, of pride, and conceit in his good looks, lacks the power to ſave them from premature decay, and total deſtruction ? Showing, that in the enjoyment of his artificial comforts and pleaſures, he deſtroys his teeth, his good looks, and often his life, in his thoughtleſs departure from natural ſimplicities and inſtinct.

The Young Readers, whom I imagine myſelf now addreſs-ing, are old enough to read my advice, and to underſtand it, and conſequently able to make, and to perſevere in, their own determined reſolutions, which will be ſure to conquer in the end, the habit alluded to, if it has already been allowed to grow upon them.

I adviſe you to turn back and read again, unleſs you can diſtinctly recollect it, the perfect ſucceſs that I met with in my own caſe, even at a far more advanced age, and conſequently the habit more difficult to correct ; and reſolve at every moment of your waking hours (except when it is *neceſſary* to open them) to keep your lips and teeth firmly preſſed together ; and your *teeth*, at all events, under any and every emotion, of pain or of pleaſure, of fear, of ſurpriſe, or admiration ; and from a con-tinual habit of this ſort, which will prepare you to meet more calmly and coolly the uſual excitements of life, you will find it extending through your ſleeping hours, if you will cloſe your lips and your eyes in the fixed determination, and effectually correcting or preventing the diſguſting and dangerous habit of ſleeping with the mouth open.

Not only manly beauty is produced, and manly firmneſs of character expreſſed by a habitual compreſſion of the lips and teeth ; but courage, ſteadineſs of the nerves, coolneſs, and power are the infallible reſults.

Men who have been joftled about amongft the viciffitudes of a long life, amidft their fellow men, will have obferved that all nervoufnefs commences in the mouth.   Men who lack the courage to meet their fellow men in phyfical combat, are afraid, not of their enemy, nor from a conviction of their own inferiority, but from the *difarming* nervoufnefs of an open and tremulous mouth ; the vibrations of which reach and weaken them, to the ends of their fingers and their toes.   In public debates—in the Forum or the Pulpit, a fimilar alarm refults in their certain defeat; and before a hive of Bees, in the fame want of confidence, the *odour* of *fear* which they emit, is fure to gain them the fting.

In one of the exciting fcenes of my roaming life, I recollect to have witneffed a ftrong illuftration of the above remarks, while refiding in one of the Sioux Villages, on the banks of the upper Miffouri.   A ferious quarrel having arifen between one of the Fur Company's men and a Sioux Brave, a challenge was given by the Indian and accepted by the White Man, who were to meet upon the prairie, in a ftate of nudity, and unattended ; and decide the affair with their knives.

A few minutes before this horrible combat was to have commenced, both parties being on the ground, and perfectly prepared, the Factor and myfelf fucceeded in bringing them to a reconciliation, and finally to a fhaking of hands; by which we had the fatisfaction of knowing, beyond a doubt, that we had been the means of faving the life of one of thefe men; and a fhort time afterwards, while alone with the Indian, I afked him if he had not felt fears of his antagonift, who appeared much his fuperior in fize and in ftrength—to which he very promptly replied—" no, not in the leaft; I never fear harm from a man

who can't fhut his mouth, no matter how large or how ftrong he
may be." I was forcibly ftruck with this reply, as well as with
the conviction I had got in my own mind (and no doubt from
the fame fymptoms) that the white man would have been killed,
if they had fought.

That there is an unnatural and lafting *contour*, as well as an
expreffion of uglinefs and lack of manly firmnefs of character produced in the human face by the habit I have
defcribed, every difcerning member of fociety is able eafily to
decide.

Natural.   Changed by habit.

No one would hefitate a moment in deciding which of thefe
he would have the moft reafon to fear in battle, or which to
choofe as his Advocate, for the protection of his life or his
property.

No young Lady would delay a moment, in faying which of
thefe, in her eftimation, is the beft looking young man; or
deciding (in her own mind) which of them fhe would prefer
for her Suitor, provided fhe were to take either.

No one would hefitate in deciding which of thefe horfes to
buy (provided the poor Brutes were victims to fuch misfortunes).

And no one, moſt aſſuredly, ſo poor a Phyſiognomiſt as not to decide in a moment, which of theſe young Ladies was the moſt happy, and which would be likely to get married the firſt;

Nature.   Habit.

and from theſe innocent and helpleſs ſtartings in life, it is eaſy to perceive how man's beſt ſucceſs, or firſt and worſt misfortunes are foreſhadowed, and the fond mother, whilſt ſhe watches, in thoughtleſs happineſs, over her ſleeping idol, may read in that little open mouth, the certain index to her future ſorrows.

It has already been faid that man is an "open mouthed animal," and alfo fhown that he is only fo by *habit*, and not by *Nature ;* and that the moft ftriking difference which is found to exift between Mankind in Savage and Civil ftates, confifts in that habit and its confequences, to be found in their relative fanitary conditions.

The American Savage often *fmiles*, but feldom *laughs ;* and he meets moft of the emotions of life, however fudden and exciting they may be, with his lips and his teeth clofed.   He is, neverthelefs, garrulous and fond of anecdote and jocular fun in his own firefide circles; but feels and expreffes his pleafure without the explofive aïtion of his mufcles, and geftiçulation, which charaïterize the more cultivated Races of his fellow men

Civilized people, who, from their educations, are more excitable, regard moft exciting, amufing, or alarming fcenes with the mouth open; as in wonder, aftonifhment, pain, pleafure, liftening, &c., and in *laughing*, draw pleafure in currents of air through their teeth, by which they infure (perhaps) pain for *themfelves*, in their fober moments, and for their *teeth*, difeafes and decay which no Dentifts can cure.

The Savage, without the change of a mufcle in his face, liftens to the rumbling of the Earthquake, or the thunder's crafh, with his hand over his mouth, and if by the extreme of other excitements he is forced to laugh or to cry, his mouth is invariably hidden in the fame manner.

As an illuftration of fome of the above remarks, perhaps " *Punch and Judy*," which is generally as apt as any other exciting fcene to unmafk the juveniles, may with effeïl be alluded to for contraft of expreffion, as familiar in our ftreets, or as it

would be viewed by an equal multitude of favage children.

It is one of the misfortunes of Civilization, that it has too
many amufing and exciting things for the mouth to fay, and

too many delicious things for it to tafte, to allow of its being clofed during the day; the mouth, therefore, has too little referve for the protection of its natural purity of expreffion; and too much expofure for the protection of its garniture; and, ("good advice is never too late") keep your mouth fhut when you *read*, when you *write*, when you *liften*, when you are in pain, when you are *walking*, when you are *running*, when you are *riding*, and, *by all means*, when you are *angry*. There is no perfon in fociety but who will find, and acknowledge, improvement in health and enjoyment, from even a *temporary* attention to this advice.

Mankind, from the caufes which have been named, are all, more or lefs invalids, from infancy to the end of their lives; and he who would make the moft of life under thefe neceffary ills; fecure his good looks, and prolong his exiftence; fhould take care that his lungs and his teeth, however much they may be from habit, or from neceffity, abufed during the day, fhould at leaft be treated with kindnefs during the night.

The habit againft which I am contending, when ftrongly contracted, I am fully aware, is a difficult one to correct; but when you think ferioufly of its importance, you will make your refolutions fo ftrong, and keep them with fuch fixed and determined perfeverance, that you will be fure to fucceed in the end.

If you charge your minds during the day fufficiently ftrong, with any event which is to happen in the middle of the night, you are fure to wake at, or near the time; and if fo, and your minds dwell, with fufficient attention, on the importance of this fubject during the day, and you clofe your eyes and your teeth at the fame time, carrying this determination into your fleep, there will be a ſtrong monitor during your reſt, that your mouth muſt be ſhut; and the benefits you will feel during the following day, from even a partial ſucceſs, will encourage you to perſevere, until at laſt, the grand and important objeᶜt will be accompliſhed.

One ſingle ſuggeſtion more, Young Readers, and you will be ready to be your own Phyſicians, your own proteᶜtors againſt the horrors of the Nightmare, Snoring, and the dangerous diſeaſes above deſcribed.

When you are in a theatre, you will obſerve that moſt perſons in the pit, looking up to the gallery, will have their mouths wide open; and thoſe in the gallery, looking down into the pit, will be as ſure to have their mouths ſhut. Then, when you lay your head upon your pillow, advance it a little forward, ſo as to imagine yourſelf looking from the Gallery of a Theatre into the Pit, and you have all the ſecrets, with thoſe before mentioned, for diſpelling from you the moſt abominable and deſtruᶜtive habit that ever attached itſelf to the human Race.

To *Men* and *Women*, of maturer age and experience, the ſame advice is tendered; but with them the habit may be more difficult to correᶜt; but with all, it is worth the trial, becauſe there is no poſſibility of its doing any harm, and it coſts nothing.

For the greater portion of the thouſands, and *tens* of thouſands of perſons ſuffering with weakneſs of lungs, with Bronchitis, Aſthma, indigeſtion, and other affeᶜtions of the Digeſtive and Reſpiratory organs, there is a *Panacea* in this advice too valuable to be diſregarded, and (generally) a relief within their own reach, if they will avail themſelves of it.

Approach the bedſides of perſons ſuffering under either of

the above dangerous difeafes, and they will be found to be
fleeping with their mouths wide open, and working their lungs
with an over-draught of air upon them, and fubject to its midnight changes of temperature as the fires go down; and thus
nightly renewing and advancing their difeafes which their Phyficians are making their daily efforts in vain to cure.

To fuch perfons my ftrongeft fympathy extends, for I have
fuffered in the fame way; and *to* them I gladly, and in full confidence of its beneficial refults, recommend the correction of the
habit, in the way I have defcribed; their ftern perfeverance in
which will foon afford them relief; and their firft night of
natural fleep will convince them of the importance of my
advice.

Man's life (in a certain fenfe) *may* be faid to " be in his own
hands," his body is always clofely invefted by difeafes and
death.   When awake, he is ftrong, and able to contend with,
and keep out his enemies; but when he is afleep he is weak;
and if the front door of his houfe be then left open, thieves and
robbers are fure to walk in.

There is no harm in my repeating that Mothers fhould be
looked to as the firft and principal correctors of this moft
deftructive of human habits; and for the cafes which efcape
their infant cares, or which commence in more advanced ftages
of life.   I have pointed out the way in which every one may
be his or her Phyfician; and the united and fimultaneous efforts
of the Civilized World fhould alfo be exerted in the overthrow
of a Monfter fo deftructive to the good looks and life of man.
Every Phyfician fhould advife his patients, and every Boarding
School in exiftence, and every hofpital, fhould have its furgeon
or matron, and every Regiment its Officer, to make their

nightly, and *hourly*, "rounds," to force a ftop to fo unnatural, difgufting, and dangerous a habit.

Under the working of fuch a fyftem, mothers guarding and helping the helplefs, Schoolmafters their fcholars, hofpital furgeons their patients, Generals their foldiers, and the reft of the world protecting themfelves, a few years would fhow the glorious refults in the Bills of Mortality, and the next generation would be a *Re-generation* of the Human Race.

The Reader will have difcovered, that in the foregoing remarks (unlike the writer of a Play or a Romance, who follows a *plan* or a *plot*) I have aimed only at jotting down, with little arrangement, fuch facts as I have gained, and obfervations I have made, in a long and laborious life; on a fubject which I have deemed of vaft importance to the human Race; and which, from a *fenfe* of *duty*, I am now tendering to my fellow beings, believing, that if fufficiently read and appreciated, thoufands and tens of thoufands of the human family may, by *their own* efforts, refcue their lives, and thofe of their children, from premature graves.

And in doing this, I take to myfelf, not only the fatisfaction of having performed a *pofitive duty*, but the *confolations*, that what I have propofed can be tried by all claffes of fociety alike, the Rich and the Poor, without pain, without medicine, and without expenfe; and alfo, that thoufands of fuffering wanderers in the wilderneffes and malaria of foreign lands, as well as of thofe in the midft of the luxuries of their own comfortabl homes, will privately thank me in their own hearts, for hints they will have got from the foregoing pages.

The Proverb, as old and unchangeable as their hills, amongft

the North American Indians: " My fon, if you would be wife,
open firft your Eyes, your Ears next, and laft of all, your
Mouth, that your words may be words of wifdom, and give no
advantage to thine adverfary," might be adopted with good
effect in Civilized life; and he who would ftrictly adhere to it,
would be fure to reap its benefits in his waking hours; and
would foon find the habit running into his hours of reft,
into which he would calmly enter; difmiffing the nervous
anxieties of the day, as he firmly clofed his teeth and his lips,
only to be opened after his eyes and his ears, in the morning;
and the reft of fuch fleep would bear him *daily* and *hourly* proof
of its value.

And if I were to endeavour to bequeathe to pofterity the
moft important Motto which human language can convey, it
fhould be in *three words*—

Shut—your—mouth.

In the focial tranfactions of life, this might have its beneficial refults, as the moft friendly, cautionary advice, or be
received as the groffeft of infults; but where I would point and
engrave it, in every *Nurfery*, and on every *Bed-poft* in the
Univerfe, its meaning could not be miftaken; and if obeyed,
its importance would foon be realized.

Geo. Catlin.

APPENDIX.

FROM the obfervations, with their refults, on board of a Mail Steamer, given in a former page, together with numerous others of a fimilar nature made whilft I have been in the midft of Yellow fever and the Cholera in the Weft India Iflands and South America; I confcientioufly advance my belief, that in any Town or City where either of thofe peftilences commences its ravages, if that portion of the inhabitants who are in the nightly habit of fleeping with their mouths open were to change their refidence to the country, the infection would foon terminate, for want of fubjects to exift upon.

This opinion may be ftartling to many; and if it be *combated*, all the better; for in fuch cafe the important experiment will more likely be made.

AUTHOR.

Rio Grande, Brazil, 1860.

Made in the USA
Middletown, DE
02 December 2023